Edward Sherman Gould

High Masonry Dams

Edward Sherman Gould

High Masonry Dams

ISBN/EAN: 9783744763950

Printed in Europe, USA, Canada, Australia, Japan

Cover: Foto ©berggeist007 / pixelio.de

More available books at **www.hansebooks.com**

HIGH

MASONRY DAMS.

BY

E. SHERMAN GOULD,

M. AM. SOC. C. E.,

CONSULTING ENGINEER FOR WATER WORKS.

AUTHOR OF "A PRIMER OF THE CALCULUS",
"ARITHMETIC OF THE STEAM ENGINE", ETC.

NEW YORK:

D. VAN NOSTRAND COMPANY,

23 MURRAY AND 27 WARREN STREETS,

1897.

TABLE OF CONTENTS.

PREFACE.

THE present volume replaces the original No. 22 of Van Nostrand's Science Series, bearing the same title, by Mr. John B. McMaster.

Mr. McMaster's volume treated mainly of the mathematical calculation of high masonry dams as it was understood at the time at which he wrote. Since then the masterly treatise of Mr. Edward Wegmann, upon the same subject, has so completely superseded all other treatment of the mathematical features involved that it would be useless to revive old methods.

Besides, the present author has long been convinced of the fact that, in view of the many practical limitations which surround the design of a high masonry dam, it is useless to attempt to adhere to a general formula for that great desideratum of all economical engineering design; namely, a SECTION OF EQUAL RESISTANCE.

The mathematical researches of those authors who have investigated this problem have established a vertical section, the basis of which is a right-angled triangle of base equal to two-thirds or three-quarters of its height, as that leading to, or at least looking towards, such a result. The most refined calculations will inevitably bring us back to the neighborhood of this form for at least the first hundred or two hundred feet of any proposed dam, the difference of relation between the base and height of the triangle depending mainly upon the limiting unit stress adopted.

We cannot do better, therefore, than to profit by, rather than to repeat, the labors of those distinguished mathematicians and engineers who have been our pioneers in this work, and start our designs by first laying down such a triangle, surmounting it by a proper practical top width instead of its own sharp apex, and, if its height exceeds 80 to 100 feet, giving a flare to the lower part of its inside face to expand the footing on that side. Then, by simple and well-known processes, we determine the maximum compressive stress upon the material at certain different heights, upon the two assumptions of an empty and a full reservoir, and if they do not prove satisfactory, modify the section accordingly, subordinating the modifications to certain practical conditions previously determined upon.

It will be seen that the dangerous stress in a very high masonry dam is the crushing one. The author has endeavored to treat the question of this particular stress quite fully, beginning with its consideration when uniformly distributed, and showing the manner in which, theoretically, a structure can be proportioned so as to render this stress uniform, no matter to what height it may be raised. In this way the rapid increase of base, after a certain height has been reached, necessary to secure this uniformity of pressure, is clearly shown, indicating that there is a practical limit to the height to which the structure can be raised. The investigation then passes to the consideration of maximum unit stresses when the resultant of pressures cuts the base unsymmetrically, as is the case in dams.

The proper practical section of a high masonry dam having been evolved, the manner of executing the work is then briefly treated of, with a description of the necessary accessories to the dam, in order that the purposes for which it was built may be satisfactorily accomplished. It is hoped that this portion of the book may prove a valuable addition to the subject.

It may be useful to the reader who cares to look farther into the general subject of dam and reservoir building, to mention that considerable additional information is to be found in the second revised edition of "The Designing and Construction of Storage Reservoirs," which forms No. 6 of the present series.

E. S. G.

Yonkers, N. Y., *July*, 1897.

CHAPTER I.

STATIC STRESSES.

LET A B C D, fig. 1, represent the vertical section of a rectangular wall of masonry sustaining the pressure of a body of water level with its top.

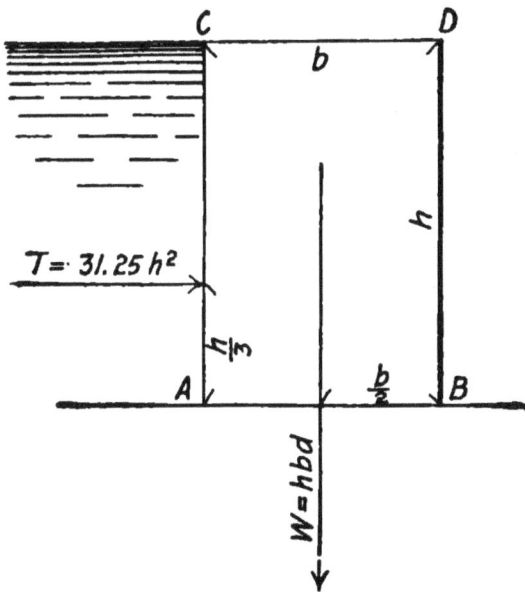

Fig. I.

The pressure of the water tends first to push the wall forward, bodily, by causing it to slide upon its base. If this tendency is resisted, the next effort of the water-pressure is to overturn the mass by causing it to rotate around the point B.

With what force does the water press, and with what strength does the wall resist?

Considering always a slice of wall and water one foot thick, so that areas in square feet will represent volumes in cubic feet, and, assuming the density of water to be 62.50 lbs. per cubic foot, and representing the height of the wall by h, then by well-known principles of hydrostatics, the horizontal thrust T * of the water against the surface A C is:

$$T = 31.25 \, h^2 \dots\dots\dots\dots\dots (1)$$

This is the pressure tending to push the wall forward upon its base. The pressure is resisted by the weight of the mass represented by the area A B C D, multi-

---

* The horizontal thrust is always expressed by (1) no matter what form the surface A C may assume, whether vertical, inclined, plane or curved, h being the depth of water pressing against it.

plied by its coefficient of friction. This
latter factor is a very uncertain quantity.
It is frequently assumed as about 0.75.
With good masonry it is probably a great
deal more, because the resistance then
becomes the transverse strength of the
masonry, or its resistance to shearing,
rather than the mere resistance of friction.

However, to conform to usage, we will
assume a coefficient of friction of 0.75.
Then, the thickness A B or C D of the
wall being represented by b, and the
density or weight of a cubic foot of the
masonry, by d, its weight will be $h\,b\,d$,
and the resistance to sliding, R of the
wall or the weight multiplied by the
coefficient of friction, will be

$$R = 0.75\,h\,b\,d \ldots\ldots\ldots\ldots\ldots\ldots (2) \cdot$$

For equilibrium:

$$31.25\,h^2 = 0.75\,h\,b\,d.$$

With a given height and density the
required factor is b, and we have

$$b = \frac{41.67\,h}{d} \ldots\ldots\ldots\ldots\ldots\ldots(3)$$

This is the equation for exact equilibrium.   Using a factor of safety of 2, the
above becomes

$$b = \frac{83.3 \; h}{d} \dots\dots\dots\dots\dots\dots\dots (4)$$

Assuming a light brick wall $d = 115$,
then, in round numbers

$$b = \frac{3}{4} h.$$

Assuming a granite or limestone wall,
$d = 165$

$$b = \frac{h}{2}.$$

The above values of b represent the
two extremes of weight of wall, and show
that, under conventional assumptions, the
height of such a wall should not exceed
from 1½ to 2 times its thickness.

Next, as regards the tendency to overturn.   The force tending to produce
overturning is the horizontal thrust T,
multiplied by the height of its point of
application above the point of rotation B.

This height, by well-known principles, is
$\frac{h}{3}$; hence the overturning moment of

the thrust M T is

$$M T = 10.42 \, h^3. \dots\dots\dots\dots(5)$$

The resisting moment, M R, of the wall is its weight multiplied by the horizontal distance of its centre of gravity from the point of rotation, in this case $= \dfrac{b}{2}$. Then

$$M R = \frac{h \, d \, b^2}{2} \dots\dots\dots\dots\dots(6)$$

For equilibrium,

$$10.42 \, h^3 = \frac{h \, d \, b^2}{2},$$

$$b = \frac{4.57 \, h}{\sqrt{d}} \dots\dots\dots\dots\dots(7)$$

Assuming a factor of safety, 2, in (6),

$$b = \frac{6.46 \, h}{\sqrt{d}} \dots\dots\dots\dots\dots(8)$$

assuming $d = 115$;

$$b = 0.60 \, h;$$

assuming $d = 165$

$$b = 0.50 \, h.$$

As against overturning, therefore, the density of the masonry has but a comparatively slight effect upon the thickness of the wall, since its square root only enters the calculation, instead of its full value, as in the resistance of the wall to sliding.

In round numbers, and for safety, we will assume always

$$b = \frac{2\,h}{3}.$$

The tendency of the wall to slide on its base is somewhat balked by the tendency to capsize, which prevents a fair and square shove taking place. Except in some special cases, where there is an abnormal tendency to slide, such as a masonry wall built upon a timber platform with the planking running in the direction of the thrust, experience shows that, if the wall is secure against overturning, it is also against sliding; so that this latter cause of failure need not be considered in general.*

Fig. 1 shows what is called a *plumb wall;* that is, one of which the vertical section is

---

* When a wall stands upon a slippery foundation, however, its stability as against sliding must be carefully considered, for in such a case, it may be much less than that against overturning.

a rectangle. Walls for dams are never built upon this section; but the conditions of stability having been determined for such a wall, its section is very easily transformed into an equivalent trapezoid. This can be done by "trial and error," but a sufficiently close approximation is arrived at by using Vauban's rule, which may be exemplified as follows:

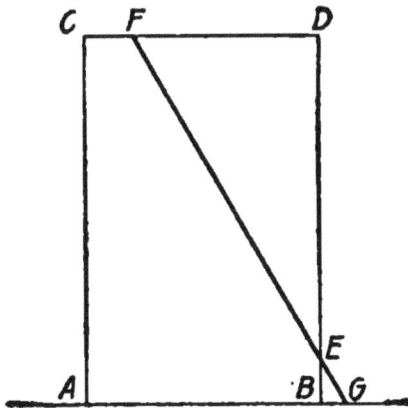

Fig. 2.

Let A B C D, fig. 2, represent a plumb wall. On B D take $BE = \dfrac{BD}{9}$. Then, through the point E, any line, F G, will form a trapezoid A G C F, of which the resisting moment will be very nearly equal to that of the rectangle A B C D. This transformation results in a very considerable economy, for the triangle E F D is replaced by the very much smaller one B E G.

To test all this by an example, referring to fig. 2, let A C = 45 ft. and A B = 30 ft.; let B E = 5 ft. and C F = 6 ft.; then, by similar triangles,

$$\frac{B\ G}{5} = \frac{24}{40},$$
$$B\ G = 3\ ft.,$$
$$and\ A\ G = 33\ ft.$$

The area of the rectangle is 1350 square feet, and that of the trapezoid 877.75 square feet, or 65 per cent. of the rectangle. The moment of resistance of the rectangular section is

$$\frac{45 \times 30 \times 30}{2} = 20250.$$

To find that of the trapezoid, divide it into a rectangle and a triangle (fig. 3). Then the moment of resistance of the rectangle around the point G is

$$45 \times 6 \times 30 = 8100 ;$$

that of the triangle is

$$\frac{45 \times 27 \times 18}{2} = 10935.$$

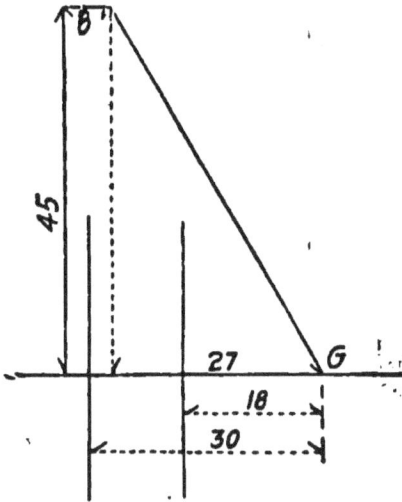

Fig. 3.

Their sum is 19035, or 94 per cent. of. the moment of resistance of the rectangle; hence 35 per cent. of economy has been gained at the expense of 6 per cent. of stability.

To test the factor of safety of above section, find first the overturning moment of water-pressure from (5),

$$M\,T = 10.42 \times \overline{45}^{3} = 949523 \text{ foot pounds.}$$

To find that of the wall, supposing its density to be 140 lbs. per cubic foot,

M R = 19035 × 140 = 2664900 ft. pds.

Factor of safety,

$$\frac{2664900}{949523} = 2.80.$$

The section is acted upon by two forces, the thrust of the water and the weight of the mass, acting at right angles to each other. They will have an oblique resultant, cutting the base at some point which may be easily determined. For this it will be first necessary to find the horizontal distance, x, of the centre of gravity of the mass from the outer toe G of the dam. Thus, referring to previous calculations,

$$x = \frac{19035}{877.75} = 21.70 \text{ feet.}$$

Now, referring to fig. 4, we have the weight of the mass 122885 lbs. acting vertically downward through the centre of gravity of the mass, at the distance already determined, 21.70 ft. from the toe. The thrust of the water, 63281 lbs., acting horizontally at the height of $\dfrac{45}{3} = 15$ ft. from the base, cuts the line of action of the

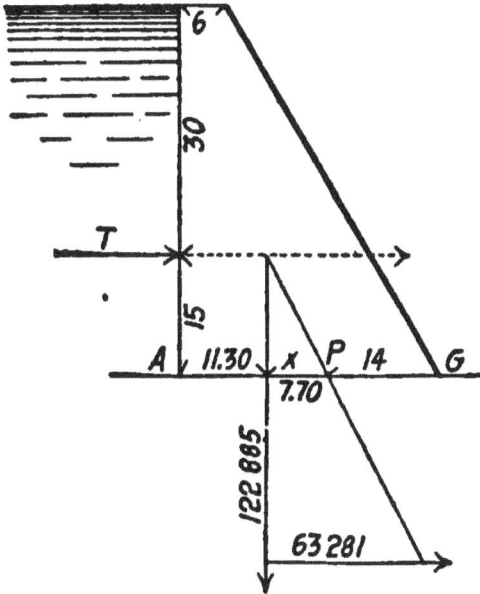

Fig. 4.

weight, and enables us to construct the triangle of forces shown in the figure.

These data are all that are necessary to calculate the distance G P of the point P, where the resultant cuts the base from the toe G; for, from similar triangles, we have

$$\frac{x}{63281} = \frac{15}{122885}.$$

$$x = 7.70 \text{ ft.}$$

Then G P = 21.70 — 7.70 = 14.00 ft.

NOTE: There are two very convenient formulæ giving the bottom width, b, of a trapezoidal wall of height, h, top width, a, and density d, when a factor of safety, f, is desired. They are, as against sliding, the coefficient of friction being c

$$b = \frac{62.50\,f\,h}{c\,d} - a \dots\dots\dots\dots(a)$$

As against overturning

$$l = \tfrac{1}{2} \sqrt{\frac{125\,f\,h^2}{d} + 3a^2} - \frac{a}{2} \dots\dots\dots(b)$$

As an example, let $b = 30$ ft.; $a = 6$ ft.; $d = 125$ lbs.; $f = 2$, and $c = 0.75$. Then, as against siding. using (a)

$$b = 34 \text{ ft.}$$

As against overturning, using (b)

$$b = 18.84 \text{ ft.}$$

This is a striking illustration of the danger of slippery foundations, as referred to in the foot note to page 12.

CHAPTER II.

UNIT STRESS.

So far two causes only of failure have
been considered, which indeed generally
reduce to one—namely, overturning around
the outer toe of the dam. In dams up to,
say, 80 feet high, this is the only case
that need be considered. Beyond this
height another element of destruction forces itself into the problem, namely, the
possible crushing of the material under
the heavy load of its own weight. It is
this consideration that places the HIGH
MASONRY DAM in a class by itself, subject
to much more complicated conditions than
its more diminutive analogue.

As resistance to overturning generally
embraces resistance to sliding, so does resistance to crushing, in a high dam, embrace that to overturning; so that, in a
high masonry dam, we need only trouble
about this cause of failure.

The problem regarding resistance to crushing consists, first in determining what degree of unit stress or pressure per square foot the material can safely sustain, and then in so designing the vertical section that all the lower portion shall be a section of *equal resistance;* that is, so that if any number of horizontal planes be passed through this portion, the maximum pressure per square foot shall be at least approximately equal, and in no case shall exceed the designated limit.

It will be necessary to elucidate this subject very thoroughly, because it contains the kernel of the whole investigation.

Let fig. 5 represent a rectangular prism of homogeneous masonry. It is evident that every square foot of the base receives the same degree of pressure, and if it were placed upon a plane surface of soft material it would imbed itself to an equal depth over the entire area of the base. If it were placed upon a hard surface, and its height were sufficient to produce a crushing stress, destruction would probably commence by a scaling off of the

edges and corners; for it is there that the
tendency to lateral expansion—the in-
variable result of severe compression—
would encounter the least resistance.

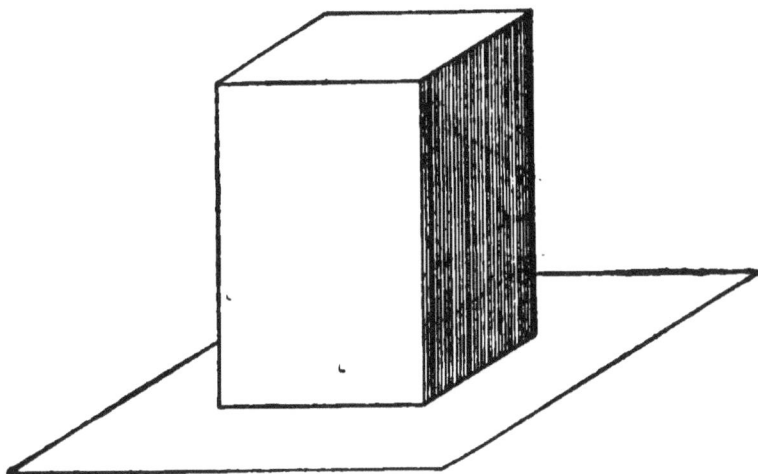

Fig. 5.

If the mass, instead of being in the form
of a rectangular prism, had that of a pyramid symmetrical about a vertical axis passing through its centre of gravity, the unit
stress, or pressure per square foot, would
also be uniform throughout its base—at
least *it is assumed so to be;* but it is evident that the pyramid could be raised a
great deal higher than the prism—three
times higher—before the crushing stress

was reached. Its corners and edges would, however, be more liable to "check" than those of the prism, because sharper.

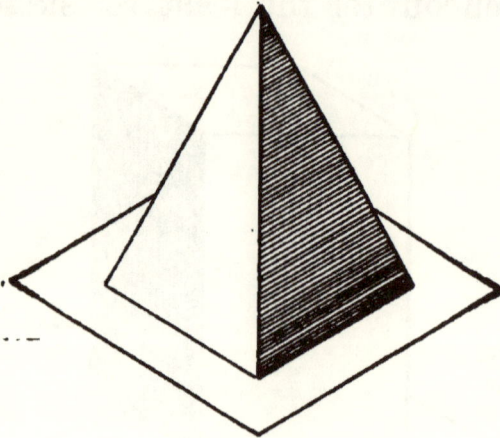

Fig. 6.

If the mass were in the form shown in

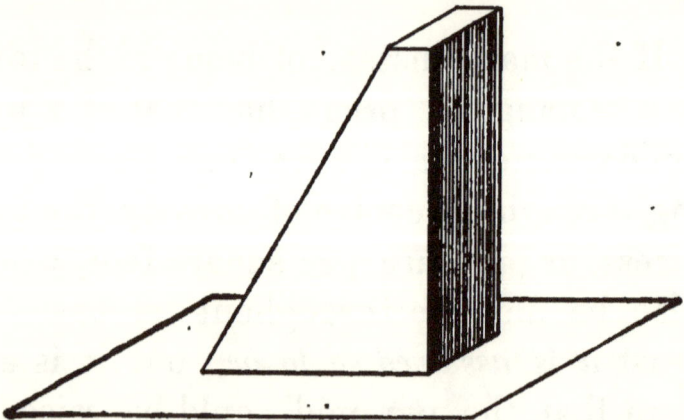

Fig. 7.

fig. 7, the unit stress would be no longer

uniform throughout the area of the base; it would be greater to the right, where the superimposed weight is greater. If the mass were placed upon a slowly yielding substance it would gradually settle to the right, and finally topple over. If placed upon a hard surface, crushing would be looked for first at the right hand side, where the pressure was greatest.

Leaving the consideration of unequally distributed stress for the moment, let us revert to the case shown in fig. 5. It is evident that the rectangular prism might be carried up to such a height as to develop the maximum permissible unit stress at the base. This would be the limiting height of the pillar of masonry. But by placing it upon a pedestal of the same material of gradually widening section, we might raise it to a much greater height, provided the ratio of expansion of the base corresponded with that of the increasing weight.

The question would then be to determine the law of correspondence between the two. This may be done by a simple algebraic process.

Let 0, fig. 8, represent the foot of a rectangular prism or obelisk of height H, and weight W, the area in square feet of its base being represented by a. Let it stand

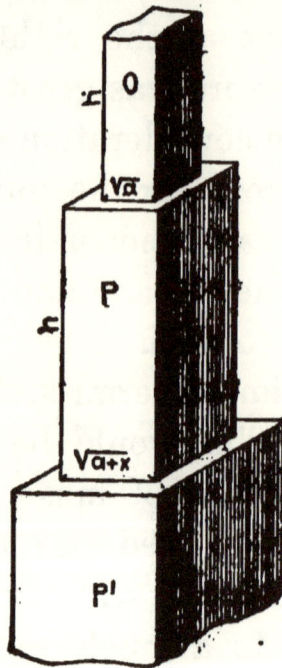

Fig. 8.

upon a pedestal P of the same material, or one of equal density d, of height h, and let it be required to find the area $a + x$, which the pedestal should have so that the unit stress upon its base may be the same as at the base of the obelisk. The additional area x must therefore bear the same relation to the weight of the pedestal that the area a does to the weight of the obelisk; that is, the weight of the pedestal being $(a + x)\, d\, h$, we must have

$$\frac{W}{a} = \frac{(a + x)\, d\, h}{x},$$

$$x = \frac{a^2\, d\, h}{W - a\, d\, h}.$$

The total area of base of pedestal, $A = a + x$, will be

$$A = \frac{W\, a}{W - a\, d\, h} \quad\dots\dots\dots\dots(9)$$

and the side of the square

$$S = \sqrt{\frac{W\, a}{W - a\, d\, h}}.$$

These formulæ will apply to any other additional pedestal, P, by making $W =$ total weight upon it, and a the area of the base of the pedestal immediately above it.

We see by (9) that the value of A will depend upon that of h, which is the inde-

26

pendent variable. We may take h as small as we please, but cannot take it so large as to make $a\,d\,h = W$.

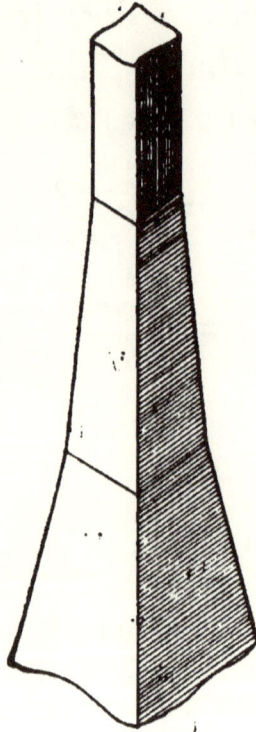

Fig. 9.

Equation (9) may be generalized by replacing W by its equivalent, $H\,a\,d$, and expressing h in terms of H; thus: $h = c\,H$, c representing any suitable fraction. We then have

$$A = \frac{H \cdot a^2 d}{H \, a \, d - a \, d \, c \, H},$$

$$\Lambda = \frac{a}{1 - c} \ldots\ldots\ldots\ldots \quad (10)$$

If the pedestal supports a statue, or any object other than the supposed obelisk, it can be reduced to an equivalent obelisk, when (9) and (10) become applicable.

As an example: Suppose the obelisk O, fig. 8, to be 100 ft. high, and its square base to contain 100 square feet. Suppose the extreme case of its being desired to raise the obelisk 1000 ft. high, by placing it on top of a series of 20 pedestals each 50 ft. high. What would be the area of the last pedestal ?

Here $c = \frac{1}{2}$, and (10) becomes

$$A = 2\, a;$$

that is, the area of each pedestal is double that of the one next above it.

The equation just given represents a geometrical progression, of which the first term is 100 and the ratio 2. The value of the 21st term, which represents the area of the 20th pedestal, is therefore 104,857,600 sq. ft., and the side of the pedestal 10,240 ft., or more than a thousand times greater than that of the obelisk.

Suppose we make $c = \dfrac{1}{100}$; that is, use 1000 pedestals each 1 foot high; then (10) becomes

$$A = \frac{100}{99} a.$$

Here we still have the geometrical progression, of which the first term is 100; but the ratio is reduced to $\dfrac{100}{99}$, and the side of the last, or 1000th pedestal becomes about 1523 ft.

It is obvious, however, that the above method of making the pedestal a series of rectangular blocks is faulty, because each pedestal is carrying an additional useless load. All the projecting offsets or steps could be advantageously removed and the blocks made in the form of truncated pyramids, as shown in fig. 9.

The weight of each would then be

$$\frac{(2\,a + x)\,d\,h}{2}$$

and the desired relation,

$$\frac{W}{a} = \frac{(2\,a + x)\,d\,h}{2\,x};$$

whence

$$x = \frac{2\,a^2\,d\,h}{2\,W - a\,d\,h},$$

and total area, A,

$$A = \frac{2\,W\,a + d\,h\,a^2}{2\,W - d\,h\,a}\ldots\ldots(11)$$

Substituting H a d and c H for W and h,

$$A = \frac{(2 + c)}{2 - c}\,a\ldots\ldots(12)$$

Let $c = \frac{1}{2}$. as before; then

$$A = \frac{5}{3}\,a.$$

The ratio of the geometrical progression is now reduced to $\frac{5}{3}$, and the side of the last pedestal drops to 1654 ft.

If we make $c = \frac{1}{100}$, then

$$A = \frac{201}{199}\,a,$$

and the side of the last, or 1000th pedestal, becomes 1485 ft.

There is much less advantage in reducing the height of the pedestal in this case than in that of the rectangular prisms.

The side of any intermediate pedestal can be found in the same way. If the truncated pyramids 50 ft. in height were used, then at a distance of 500 ft. from the top, the side of the square base of the 10th pedestal would be 128.6 ft.*

Although this example seems foreign to the subject of dam-building, yet it manifests most strikingly that, while it is theoretically possible to design a structure of equal compressive resistance of any height, yet when certain moderate heights are exceeded, the geometrical progression runs us into preclusive dimensions, recalling the famous problem of the horseshoe nails.

---

* By the aid of logarithms these calculations are rapidly and easily performed.

CHAPTER III.

UNEQUALLY DISTRIBUTED UNIT STRESS.

PREVIOUS calculations relating to fig. 4 show that the resultant weight of the trapezoidal wall, when there is no water pressure against it, acts vertically downward on a line passing through its centre of gravity and cutting the base at a point distant 11.30 ft. from the inner toe, A. When the wall sustains a full head of water, the horizontal thrust, combined with the weight of the wall, produces an oblique resultant which cuts the base 14 feet from the outer toe, G.

These two pressures, the one acting vertically downward when there is no water behind the wall, and the other obliquely downward and outward when there is a full head behind it, are essentially different in character. The first produces compression only; the second, compression and shearing stress.

This second pressure, namely, the resultant of a vertical downward force and a
horizontal thrust, may be resolved at the
point P back to its components, producing
a downward vertical pressure exactly
equal to the weight of the dam, and a
horizontal one exactly equal to the thrust
of the water. The first produces the
crushing stress, and the second the shearing stress already mentioned.

The horizontal, or shearing stress, is
rightly or wrongly wholly ignored, and
the vertical component, or crushing stress,
only considered. The water pressure
is therefore supposed to have no other
effect than to move the point of the application, or point of mean pressure of
the weight of the dam, from its position
nearer the point A to a position nearer
the point G.

Now it has already been recognized
(fig. 7) that, when the point of application
of the weight of the mass does not cut
the middle of the base, the uniformity of
the distribution of the stress is destroyed,
being intensified between such point of

application and the nearer toe of the wall.
The unit stress can therefore no longer be
determined by simply dividing total weight
by total base; and if our structure is so
high as to inspire fears of its crushing by
its own weight, we may be uncertain how
nearly it has approached the limit of safety. We know, for instance (fig. 4), that
the pressure is intensified with an empty
reservoir in the neighborhood of A, and
with a full one in the neighborhood of G;
but by how much?

On this point we have no certain knowl-
edge; but the following formulæ of French
origin, partly rational and partly empirical, have received general acceptance:

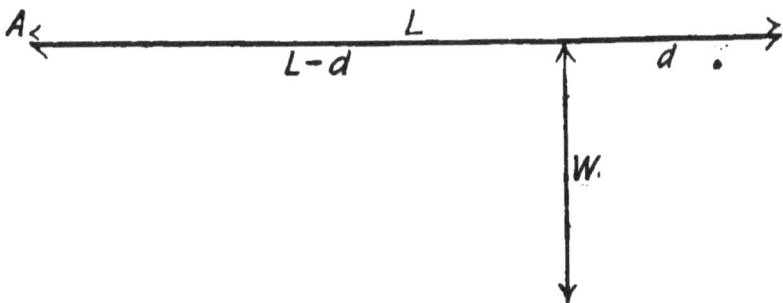

Fig. 10.

Let A B (fig. 10) represent any horizontal course of a mass of masonry of length

34

L, the resultant pressure upon which cuts
the course at the distance d from the
nearer extremity B. At this point it is resolved, if oblique, into two components,
one of which is the vertical downward
pressure, W. The two quasi empirical
formulæ giving the intensity of unit stress
at B, are then:

$$P = \frac{4\,W}{L^2}\,(L - 1.5\,d). \quad \ldots\ldots(13)$$

$$P = \frac{2\,W}{3\,d} \quad \ldots\ldots\ldots\ldots\ldots\ldots(14)$$

The first is to be used when d is equal
to, or greater than $\frac{L}{3}$, and the second when
d is equal to, or less than $\frac{L}{3}$. When $d =$
$\frac{L}{3}$, they give identical results.

The demonstration of these formulæ,
for which reference may be had to Mr.
Wegmann's "Design and Construction of
Masonry Dams," and also to an article in
the number for August, 1884, of Van Nostrand's Engineering Magazine, is far from

satisfactory; but it may be admitted that
(13) always gives safe results, and that
(14) can be confidently used when d is
only slightly less than $\dfrac{L}{3}$. In relatively
small values of d, it cannot be trusted.

An examination of these two formulæ
does not increase our confidence in their
scientific correctness. It seems odd, for
instance, since P depends upon the relative value of the variable d, that the character of the two formulæ should so greatly
and so abruptly change at the point where
$d = \dfrac{L}{3}$. The reasons for this become partly apparent on studying the manner in
which the formulæ were derived, but the
process of reasoning is by no means fully
satisfactory.

The present writer is disposed to believe
that a general and more correct formula
is the following:
$$P = \frac{W}{L\,d}\,(L-d)\ldots\ldots\ldots(15)$$
and that the pressure, as thus expressed,
extends over the entire space d, instead

of increasing from the point of application
of the resultant to the extremity B, and
becoming maximum at B, as is assumed
in the French theory. No doubt the tendency to rupture will be greater at this
point, for reasons already stated; but it is
believed that the intensity of the stress is
uniform throughout d.

The reasons leading to the adoption of
(15) are as follows: We know, or assume,
that when the resultant cuts the base in the
middle the unit stress is equal throughout the two equal segments into which it
divides the base. We know also that, as
the resultant moves away from this central
position, it intensifies the stress upon the
shorter segment, and since the total pressure upon both segments must be equal to
the total weight, that it relieves, proportionally, the stress upon the longer one.
The question is, in what proportion does
the intensity of the stress increase on one
segment and diminish upon the other?
It appears rational to assume, in view of
the above facts or admitted assumptions,
that the ratio should be *inversely as the*

lengths of the segments to the total length.
Hence, weight W^1 upon shorter segment,
the only one we are concerned about, is

$$\frac{W^1}{W} = \frac{L - d}{L},$$

$$W^1 = \frac{W\,(L - d}{L},$$

and the unit stress in d

$$P = \frac{W\,(L - d)}{L\,d},$$

as already given.

To test the agreement of the three formulæ we can express d in various fractional
values of L. For $d = \dfrac{L}{2}$, (13) and (15)
give identical results. For $d = \dfrac{L}{3}$, all three
give identical results. Between these two
limits (15) gives values a little less than
(13), the maximum divergence occurring
for a point midway between the two, or d
$= \dfrac{5}{12}\,L$, when the value of P per (15) is
$93\frac{2}{3}$ per cent. of that per (13).

For values of d less than $\dfrac{L}{3}$ (15) gives values of P rapidly increasing over those afforded by (14). When we recall that but small confidence can be placed in this latter when d differs greatly from $\dfrac{L}{3}$, and that it is precisely for such values of d that the most dangerous stresses occur, it seems clear that we should give preference to the formula offering the larger factor of safety. The limit of application of these two formulæ, (14) and (15), occurs when $d = 0$, when the mass above the course A B is on the point overturning around B. This condition is approximated if we make d = unity, taking the unit very small; or, what is the same thing, supposing the corresponding value of L to be very large. The entire mass of weight, W, is then just about to be balanced upon the very small surface $d = 1$, and yet (14) in this case would give only two-thirds of the weight, as bearing or about to bear upon the unit of surface, while (15) would give, practically, $P = W$, as is undoubtedly correct.

Since, therefore, the new formula gives total pressures which are certainly correct for the two limiting values of d, namely, $\frac{L}{2}$ and o, and agrees with both the other formulæ for $d = \frac{L}{3}$, it would seem to have much to recommend it as a safe guide for intermediate values.

It remains now to say a few words respecting the degree of crushing stress to which it is proper to submit stone masonry.

Stone masonry is a composite material, composed of stone or brick and mortar. The mortar is itself composite, being composed of sand and cement, the latter either natural or Portland, in various proportions. It is evident that in estimating the strength of such masonry we must base our calculations upon that of the weakest component. Mortar made with best imported Portland may or may not be as strong as, or even stronger than the stones which it binds together. The ultimate crushing strength of all the materials used in such structures as high masonry dams is very great; but it is evident that the loads under which carefully tested samples give way must not be even remotely approached in practice. As regards high masonry dams, the utmost conservatism should be observed in fixing the data upon which their dimensions depend, in view of the terrible consequences of failure; of the fact that such structures must be planned for a more permanent duration than perhaps any other of the works of man, and of the difficulty or impossibility of subsequently correcting any errors which may have been committed, either in their design or execution.

These considerations, together with the uncertainty which exists regarding the actual strength of the materials used, lead to the adoption of a limiting unit stress of from 15,000 to 20,000 lbs. per sq. foot, always inclining to the lower figure, which limit is entirely safe for any good hydraulic masonry.

CHAPTER IV.

The Vertical Section of High Masonry Dams.

THE French engineers were the pioneers
of scientific designing of high masonry
dams, and the noble structure across the
Furens, near St. Étienne, is the prototype
of its class.

The problem—successfully worked out—
was to ascertain by calculation the section
of *equal resistance* against crushing of such
structures, both for a full and an empty
reservoir.

It would be foreign to the purpose of
the present little volume to enter into the
mathematics of this problem, more particularly as it would be merely an interesting
study, without any real practical utility.
It is quite unnecessary to repeat these calculations each time that it is desired to
design a high dam, because a standard
type has been evolved from practical and

theoretical considerations, which we know cannot be materially varied. We may, therefore—given the height of a proposed dam—immediately lay down on paper a section to which we know it must very closely conform, and then by a few simple processes test it at certain critical points, introducing the probable density of the class of masonry which we are about to employ, and giving due weight to any other special details of the particular project under consideration.

It will be found, too, that though we may very nearly, or even exactly, obtain a section of equal resistance upon paper, the design so obtained will generally be practically inapplicable, owing to constructional requirements, which will necessitate more or less of a compromise between the ideal and the possible.

The whole subject will be best elucidated by a numerical example. Let us, therefore, proceed to design a dam 250 feet high, assuming a density of masonry of 140 lbs. per cubic foot. The completed section with full dimensions is given in

fig. 11. The maximum unit stresses are
also shown at different heights. The
manner in which these dimensions and

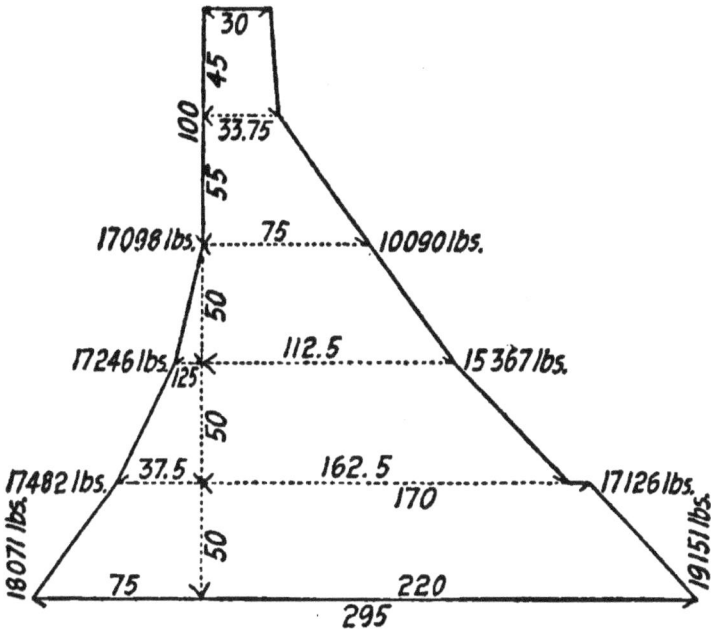

Fig. 11.

stresses have been arrived at will be best
understood by considering first the upper
one hundred feet, and working our way
down, say 50 feet at a time, until we reach
the full height of 250 feet; for high masonry dams are best built, at least on
paper, like the Irishman's chimney, by

beginning at the top and building downward.

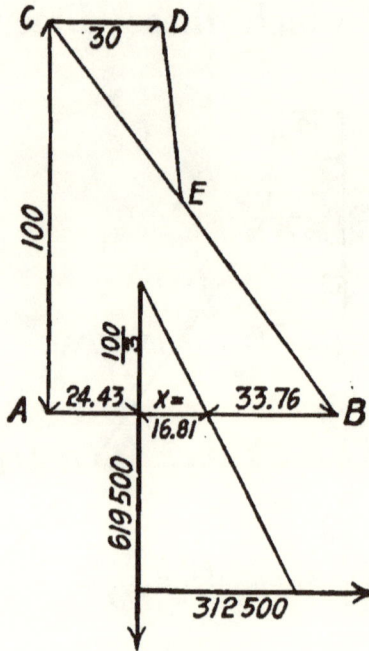

Fig. 12.

Referring, therefore, to fig. 12, we see
that this section is composed of two triangles, one A B C, right angled, of height A
C = 100 ft., and base A B = $^3/_4$ A C =
75 ft. The second, C D E, has an invert-
ed base of 30 ft., which represents the top
width of the dam. This latter dimension
is fixed arbitrarily, according to the requirements of each particular case. In

the figure it is given a face batter, D E, of
one inch to the foot, at which rate it will
intersect the slope B C of 1 to 0.75 at a
vertical distance from the top equal to one
and one half times the width C D; in this
case at a distance of 45 feet.

Our study of this section consists in as-
certaining the maximum unit stresses in
the base A B, both when the reservoir is
full and empty.   To do this it is first necessary to determine the point at which a
vertical line, drawn through the centre of
gravity of the area A C D E B, will cut the
base A B. We must first find the distance
of the centres of gravity of the two triangles from some given "axis of moments",
and then combine their moments about this
axis.   It is best, in studying such a project as that now before us, to take this axis
well to one side of the figure, so that if it
becomes desirable to modify the section as
we go on, by widening the base, all calculations already made may be still available.
Let us refer everything to an axis of moments parallel to A C, and situated 230 feet
to the right of it.

This thrust is applied at the height $\frac{100}{3}$ ft. above A B. Lay off the work as shown in fig. 12, to scale, if the operations are to be performed graphically; if by calculation, a simple sketch irrespective of scale suffices. If the work is done by calculation we require the distance x. This is obtained by similar triangles

$$\frac{3\,x}{100} = \frac{3125}{6195}\,;\, x = 16.81.$$

The distance from point of intersection of resultant to B is found, by subtraction, to be 33.76 ft. Applying (15) we find the maximum unit stress for a full dam to be 10,090 lbs.

If fig. 12 represented a dam 100 ft. high, instead of only the upper part of one a great deal higher, it would be regarded as a rather bad design; for there would be too great a difference between the maximum stresses when the reservoir was alternately full and empty. To rectify this, the base A B might be somewhat shortened to the right of A, and the difference added to the left, giving the side A C a batter from about 80 ft. below the top to the end of the extended base. Or, if proper material were at hand, an earthen embankment might be placed on the inside, against the vertical back of the dam, extending perhaps half way to the top, which would bring a permanent counter-pressure to bear, thereby reducing the unit stress when the reservoir was empty. In dams with vertical backs this earth embankment is, in the author's opinion, under all circumstances, a recommendable feature. The reason why this design is suitable for the upper part of a higher dam is, as we shall presently see, that with the increased height the pressures due to a full reservoir rapidly increase, and we must save all the extension of base possible on the inside, so as to add it to the lower side, as we keep adding to the height of our dam.

Continuing, we get the section shown in fig. 13 for the upper 150 feet of our dam. We still have the triangle, with base equal to three quarters of the height as the basis of the design, but admonished by the comparatively high unit

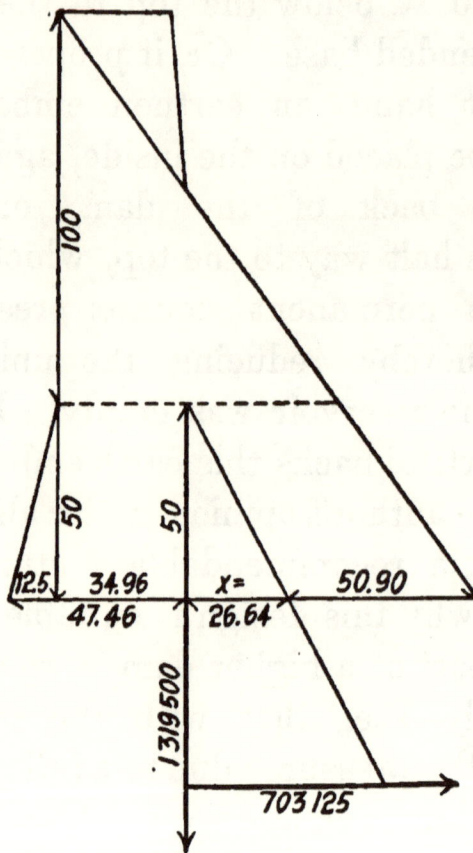

Fig. 13.

stress on the left hand side of the base when the reservoir is empty, a batter has been given to the inside face of 1 to 4, for the last 50 ft. of height. The area of the little triangle thus added is 312.50 square ft., and the distance of its centre

of gravity from the axis of moments is
234.17 ft. The area of the main triangle
is 8437.5 sq. ft. and its distance from the
axis 192.5. The elements of the upper
small triangle are the same as before.
Hence

675 .... 140906.25

8437.5 × 192.50 .... 1624218.75

312.5 × 234.17 .... 73178.13

9425.0   1838303.13

and

$$\frac{1838303.13}{9425} = 195.04 \text{ ft.}$$

The shorter segment, when the reservoir is empty, is therefore 230 + 12.50 —
195.04 = 47.46 ft., and the weight of
the mass 9425 × 140 =: 1,319,500 lbs.
The unit stress on the left-hand side of
the base is, therefore, per (15), 17, 246 lbs.
per sq. ft.

The thrust of the water, when the reservoir is full, being 703,125 lbs., we obtain,
by the same process as already employed
for the first section, a length of 50.90 ft. for
the shorter segment, and a unit stress,

when the reservoir is full, upon the right-hand side of the base, of 15,367 lbs. per square ft. This is a favorable result; for it is considered good practice to favor the outside pressures, or those corresponding to a full reservoir, rather than those corresponding to an empty one.

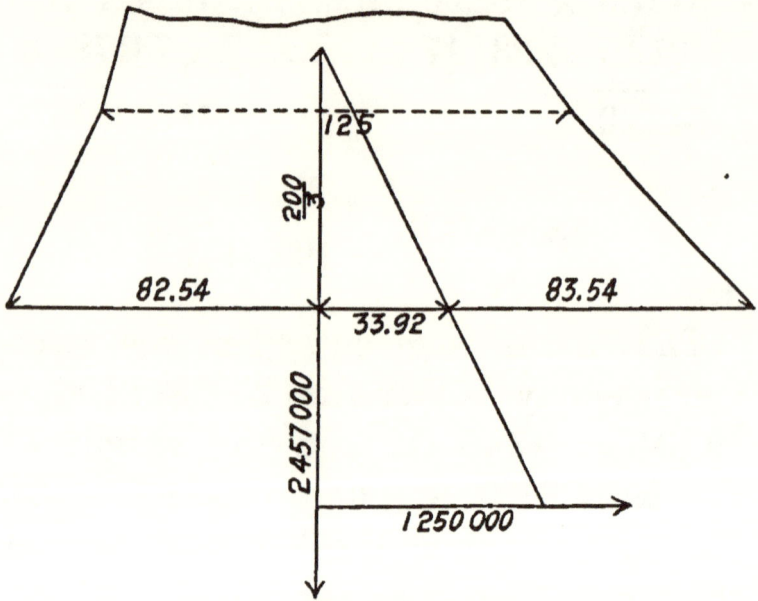

Fig. 14.

In the next addition of 50 ft., when the dam reaches the height of 200 ft., some modifications are necessary. The base must widen more rapidly, for the pressures are beginning to increase faster than

the length of base. The outside slope
is therefore increased to 1 to 1, and the
inside to ½ to 1, as shown in fig. 14.

An examination of this figure will show
that there have been added two triangles,
one of 625 square feet and the other of
1250 square feet, and a rectangle of 6250
square feet to the previous section. As-
certaining the relative positions of their
centres of gravity, and utilizing the calculations already effected, we have:

9425......................	1838303.13
625 × 250.83.........	156768.75
6250 × 180.00	1125000.00
1250 × 100.83	126037.50
17550	3246109.38

and

$$\frac{3246109.38}{17550} = 184.96 \text{ ft.}$$

Proceeding as before, we get the lengths
of d for the two assumptions of a full and
an empty reservoir as shown in the figure,
with unit stresses in the two cases of
17,126 and 17,482 lbs. per square foot.

Our increasing unit stresses admonish us that we still are not expanding our base, particularly to the right, sufficiently rapidly; for we are approaching a height where, as in the case of the obelisk, our section of equal, or approximately equal, resistance demands a more pronounced flare. Our difficulties are beginning to thicken, for good construction demands that no slope shall exceed 1 to 1, because we must at all hazards avoid expansion purchased at the cost of sharp edges, since that would only give us the appearance of a sufficiently wide base, which the feeble resisting power of the edges would render practically useless.

In our final section, for the full height of 250 ft. we must therefore have recourse to offsetting, as shown in fig. (15), and also in fig. (11), for the outside lines, and we will also flatten the inside slope down to 1 to $\frac{3}{4}$.

Proceeding exactly as before, and utilizing all previous work, we get the data shown in fig. (15), and from them deduce the unit

stress of 19,151 lbs. for a full reservoir,
and 18,071 lbs. for an empty one.

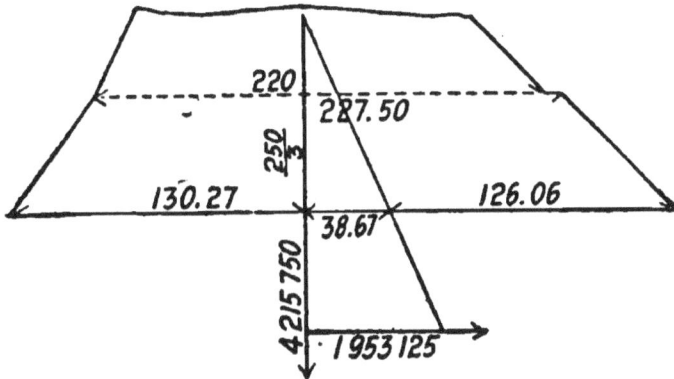

Fig. 15.

It will be seen that it has been practically impossible to adhere to a section of
equal resistance, and the best we can do
is not to exceed a given limit, and to
approach it as slowly as possible. Our
limit of 15,000 to 20,000 lbs. per square,
ft. may be regarded as a conservative one
for good masonry; for it has often been
exceeded, apparently with impunity, in actual structures. The question of what the
ultimate allowable stress is, depends upon
so many considerations, among others the
temperament of the designing engineer,
that it cannot be reduced to one of mathe-

matics. It must be constantly borne in mind that dams, both high and low, belong to a class of engineering structures against which many disastrous failures stand recorded, and we should not allow ourselves to be carried away by a desire to design a "bold" structure. Boldness in such cases may easily degenerate into recklessness, perhaps the worst characteristic short of dishonesty that a hydraulic engineer can possess.

On looking at the section shown in fig. 11, it appears indeed to be superabundantly strong, because one instinctively judges its dimensions according to their mutual proportions, and in relation to the resistance of the whole to overthrow only. This, however, is not the element of danger that we are figuring against, but just simple crushing by great and unsymmetrically disposed pressure, and the eye can guide us but little in judging of the suitableness of our design in this respect. Proportions, irrespective of actual height and consequent weight, do not affect this question; for we cannot say, of a section

adapted to one height, that it can be made
applicable to another height by simply
changing the scale of the drawing, for the
sections proper for the two cases are not
similar figures. The best way to design
a high masonry dam is to proceed as we
have done, starting, as a ground work,
with a right-angled triangle of one of
height to three quarters of base or of one
of height to two thirds of base, as in the
additional example presently to be given,
and expanding both ways as we go down,
never exceeding a slope inside or outside
of 1 to 1, and resorting to stepping when
this slope does not open out rapidly
enough.

What precedes shows us that stone masonry being the material employed, there
is a limit to the height of a dam which in
practice cannot be exceeded. The limit
has been pretty nearly reached in our example, and it would require very strong
reasons to make it advisable to exceed
this height. Dams are generally built
with the intention of forming reservoirs
for the purpose of storage. It would be

an exceptional case which did not permit
of the construction of several dams of
moderate height and equivalent aggregate
capacity rather than one of extravagant
dimensions. While up to a certain elevation the single high dam would be cheap-
er than two or more lower ones, beyond
that height the reverse would be the case.

In the preceding design for a dam 250
ft. high, it is understood that the entire
height is above ground. Nearly always,
in a high masonry dam, it is necessary to
go down a considerable distance to find a
satisfactory rock foundation, upon which
alone a purely masonry dam should be
built. This matter of foundation is apt
to be a very expensive one, on account
of the enormous volume of excavation
and of masonry which deep foundations
involve.

Fortunately, the masonry imbedded in a
deep foundation can sustain a much greater pressure than if standing up in an
isolated block, because the pressure
against the sides prevents, to a very considerable extent, the lateral expansion of

the material which is an accompaniment
of crushing. This pressure acts like the
iron band placed around the head of a
pile to prevent brooming or splitting under
the blows of the hammer.

We may adopt a maximum-unit stress
for imbedded masonry of about 30,000
lbs. per square foot.

Supposing, now, that in our example we
were obliged to go 100 ft. deep for a
proper foundation. We will let our superstructure, as already designed, rest upon a
block, rectangular in vertical section, and
projecting 20 ft. beyond the outer toe, and
15 ft. beyond the inner one, as shown in
figure 16.

Combining the moment of this mass
with the moments of the two resultants of
the superstructure corresponding respect-
ively to a full and empty reservoir, we get
the two values of $d = 155.9$ and $d = 155.6$,
shown in the figure, which give a unit
stress as above, under the two conditions
of water-pressure and absence of the same.

The object of keeping the sides of the
foundation-block vertical is that it may

receive the full advantage of the lateral support of the earth or soft rock in which it may be imbedded.

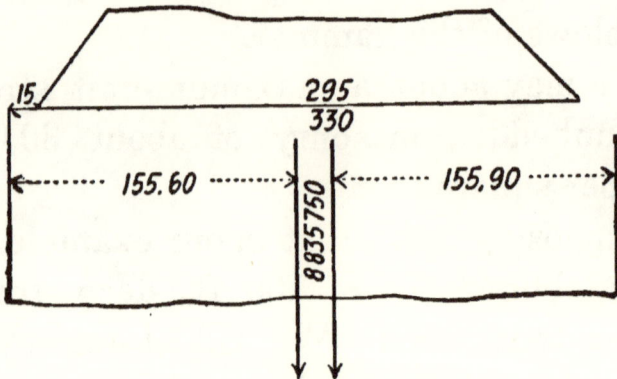

Fig. 16.

It is evident, from our calculations, that the Leavier the masonry of which the dam is composed the more favorable it is for resisting water-pressure, and the more unfavorable when resisting only its own weight.

When the dam is full, there is a certain amount of downward pressure brought upon it from the weight of the prism of water resting upon its inside polygon. This has been neglected in our calculations, because its action is somewhat uncertain. Whatever it may be, it lessens the unit stress to a greater or less degree, and constitutes a small additional factor of safety.

It remains to mention two other factors in the calculation of dams which have been neglected, but which other authors dwell upon. One is, the possible diminution of the weight of the structure, from the supposed upward pressure of water penetrating to its bottom through fissures in the rock upon which it stands. The author finds it impossible to conceive that such an upward pressure should exist to the extent of exercising any appreciable effect upon the stability of a dam, particularly one seated upon deep foundations. He considers such a factor as wholly negligible.

Nor can he agree with those who maintain that the thrust of the water from a full reservoir should be considered as that due to a head extending from the top of the dam to the bottom of the foundations. That portion of the dam which is buried in the earth or rock should, in his opinion, be considered entirely apart from the dam proper, and as subject to an entirely different class of stress. He would consider this portion of the structure as forming, in fact, a part of the geology of the territory, and confine his calculations as regards the thrust of the water to the superstructure which, standing in relief above the surface of the surrounding ground, receives the pressure of the water on one side and that of the atmosphere only on the other.

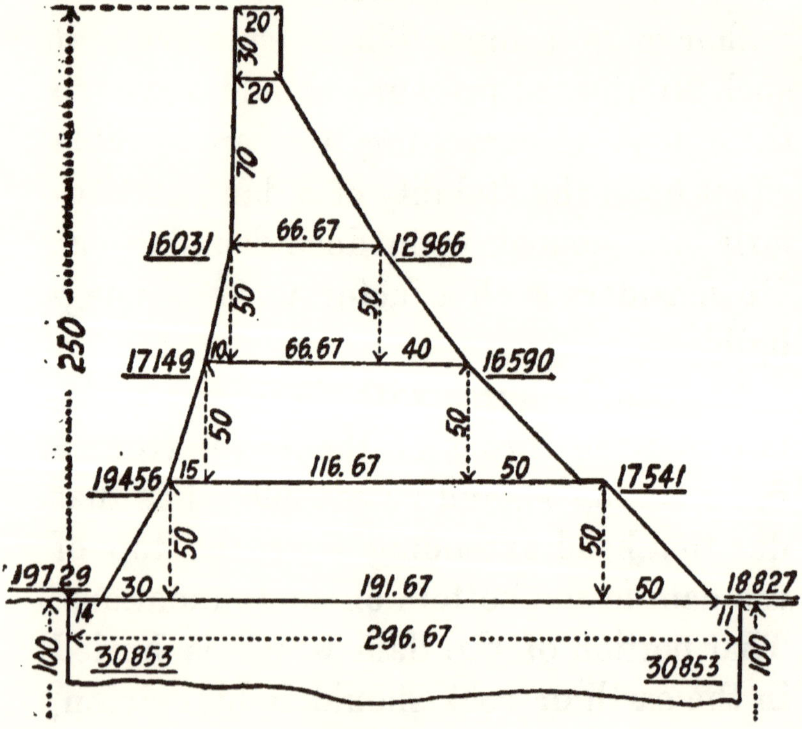

Fig. 17.

Fig. 17 is given as an additional example of a dam with the same data as before, except a slight difference at the top, and excepting also, that as a basis for the first hundred feet from the top a right angled triangle of one of height to two-thirds of base has been assumed. In this example, the stresses are but slightly in excess of those previously calculated and are rather better distributed, while the widths are materially lessened. With the given density of 140 lbs. this section is preferable to the one already given.

CHAPTER V.

THE CONSTRUCTION OF HIGH MASONRY DAMS.

THE preceding pages have treated of what may be considered as the preliminary and easiest part of the construction of high masonry dams; namely, the mathematical part. The reasoning that has been pursued has demonstrated that the sections shown in figs. 11 and 17 are perfectly safe and satisfactory, from a conservative standpoint, providing amply but not extravagantly for the stresses which our best information upon the subject leads us to expect. But it is one thing to set up a proper design on paper and quite another to successfully carry out the same on the ground. In what follows some attempt will be made to treat of the difficulties attending the execution of such structures, and the precautions to be taken in order to secure good results. Much that will be said will be found applicable to all masonry dams of whatever height; but, as all the difficulties and conditions are intensified with increased height, it will be understood generally that dams of one hundred feet and over are those under consideration.

Before proceeding with this part of the subject, a few words must first be said regarding a matter which really belongs more properly to design than construction, namely, the form of the *plan* of the dam as distinguished from its vertical section, already discussed. Should the dam be straight or curved in plan?

Much has been written on this subject, and both forms have found able advocates. Among other authorities, the "Engineering News" of New York strongly favors a curved plan, with the convexity of course directed up stream, pointing out, it is believed, for the first time, that, contrary to the received opinion a moderate degree of curvature does not increase the volume of masonry in the dam.

There can be no doubt that such a form is an element of additional strength, particularly if the dam is built across a narrow ravine, when the arch principle involved in a curved plan distinctly asserts itself. It continues to be an element of strength even in the case of a dam of such considerable length that we can no longer consider it as acting as an arch, because, since there must always be some slight forward motion in the dam when the full head of water comes against it, the effect of such motion will be to produce a compressive stress if the dam be curved, as it will tend to crowd the material into a smaller space, whereas, if the dam be built upon a perfectly straight line, the least forward motion will have the effect of opening the joints. Such motion, however, will always be so insignificant that it certainly seems an over refinement to take it into account and provide against it by the form of the dam.

On the other hand any deviation from straight lines in the ground plan of the dam involves considerable trouble and expense in laying out and executing the work, and therefore increases the difficulty

in securing good results.   Except in the
case of a short dam, as already mentioned,
the author is of the opinion that the ad-
vantages, more or less theoretical, which
are secured by a curved ground plan, are
overbalanced by the practical difficulties
which spring from it.

Very high dams are generally built
across the valleys of powerful streams,
and perhaps the greatest difficulty which
accompanies their construction is the
management of the flow of the stream
while the dam is in construction.   In the
case of small dams across relatively in-
significent brooks this difficulty is generally easily overcome, by first putting in
the pipes or gates by means of which the
water is to be drawn from the dam when
completed, and then turning the flow of the
stream through these by means of a tem-
porary dam when, profiting also by the
dry season, the rest of the foundations
can be got in without further trouble.
But generally speaking, with high dams,
deep foundations and large streams, such
a provision for the flow of the water would

be totally inadequate, and the management
of the water in such cases constitutes in
itself an engineering problem of the first
magnitude. A tunnel may be sometimes
driven entirely around the site of the dam
through which the river is turned and
completely got rid of, the difficulty then
being to properly close the tunnel, either
permanantly or by means of gates, when
the work is completed.

Perhaps the most usual way of handling
the water is to carry it in a large flume
directly over and across the foundation
pit. Very considerable streams can often
be successfully disposed of in this way
until the foundations have been got in
level with the ground, when gaps can
be left in the wall as it is carried up, and
the water shifted over from one to the
other side till the dam is finally completed.
The danger of this method is, that an un-
usual freshet may break or destroy the
flume, when the foundation pit will be
flooded, and the banks probably cave in.
It is evident also, that in the case of a
very wide excavation the great length of

the flume would preclude its use, and a
diversion of the stream will generally be
found to be the only satisfactory solution
of this most difficult problem.

Another embarrassing feature is the
manner of executing such a deep and wide
excavation as is contemplated in our
example. Shall the sides be sloped, or
shall they be timbered? It will be readily
conceded that this question does not
admit of a general answer, and would
have to be settled by the particular conditions of the individual case. The expense of securely protecting the sides of
such an excavation is so enormous, even
under the most favorable conditions, that
the tendency will always be minimize the
outlay by taking some chances. At the
same time the caving in of such a deep
and wide excavation, should it occur, involves so great a loss, not only of money
but of time, that it is poor policy to scamp
the precautions taken at the start. It is
probably safe to say that one of the prin-
cipal causes of loss incurred by contractors
in the execution of work, is to be found

in the necessity of digging out a second time excavations which have fallen in from want of proper sloping or shoring, and this is equally true proportionally, of small as well as of great undertakings. In the special case of very wide excavations, the decision will probably be to slope the sides in preference to attempting any method of timbering or shoring.

Equally important is it to provide ample pumping power when executing the excavations. Lack of this is another fruitful source of loss in time and money to the contractor. The quantity of water to be encountered, and the difficulty, as well as the necessity of getting rid of it both when taking out excavations and putting in foundations are matters almost invariably underrated when work is commenced. A most liberal allowance should be made for the quantity of water to be lifted, and the pumping sumpts should be at all times kept well below the level of the work which is going on.

So far, three important points have been considered, in their natural order, namely,

the disposition of the stream across which
the dam is to be built, both as regards
its natural and freshet flow; the manner
of executing the excavations for foundations
so as to prevent caving in of the sides, and
the means of keeping these excavations
dry, first while taking out the material,
and secondly when putting in the foundations. As regards these points, very little
more than their mere enumeration has
been attempted, although many pages
might easily be devoted to any one of
them. No written discussion would however be of much avail as throwing light
upon the subject, the only really useful
literature bearing upon these points, being
the records and reports of the actual execution of such works. Of these, the
voluminous published documents of the
Croton Aqueduct Department, and the
New Aqueduct Commission stand easily
among the best which the public works of
this country afford, but the judicious engineer, charged with the responsibility of
designing and executing any large hydraulic work of this character, will not

limit his reading and investigations to any particular school of practice, but will familiarize himself with all the varying examples within his reach, both of success and failure, in his own and other countries.

A point which now comes up is, of what class of masonry shall our dam be built? Here again we have a question not admitting of a general answer. We might indeed without much difficulty decide which among the building stones furnished by nature was the hardest, toughest, heaviest and soundest and, altogether, the best adapted for dam building; also, whether upon the whole it was best to build "Cyclopean rubble," as for the Vyrnwy dam, or go to the other extreme and use concrete. Also, we might arrive at certain similar deductions regarding the sand, cement, and proportions of mixing. But our work may be situated in such a locality that it would be impossible to procure just what we wanted, and we should be compelled to realize that after all we were reduced to the necessity of cutting our coat acccording to the cloth.

Setting aside the character and quality of the materials, and considering only the manner of putting them together, the present author is strongly inclined to believe that concrete between stone walls, rubble for the foundation and cut stone for the superstructure, would be the ideal combination of stone, sand and cement for a high masonry dam, if the proper quality of concrete as regards mixing and placing could be secured. While concrete is lighter than stone masonry, since it contains more mortar and less stone per cubic yard, and while the mixture is less strong generally speaking, than the stone of which it, or the stone masonry which it may replace, is composed, the fact of its forming a homogeneous, monolithic mass constitutes an element of superiority with which no hand placed stone work can compete. In order however to secure uniform good quality for the concrete when used upon such a gigantic scale as would be the case in these instances, a very large special plant would be required, and the strictest inspection practised from beginning to end of the work. An army of tampers, for one thing, would be required, divided into small gangs with an expert foreman over each, and numerous inspectors constantly patrolling the work. If the proper outfit for the work is secured, and a proper organization maintained, then in the author's opinion this material would be the best for such work, when confined within stone facings.

It may be suggested that the same precautions and care are necessary for good stone masonry as for good concrete, but most engineers will agree that it is easier to inspect and control stone masonry than concrete, under ordinary circumstances. One reason no doubt is that there is more skilled labor employed in laying up masonry, whether rubble or cut stone, than in putting in concrete, which is always easier to direct, and also, since the quality of concrete depends so much upon the manipulation to which it is subjected, more individual judgment is called for on the part of foremen and inspectors, than in the case of stone masonry which, when the work is once fairly started, and the requirements understood by all hands, becomes more a matter of routine.

Whatever preference may be given to one material rather than another, whether as determined by local conditions or individual prejudice or predeliction, one thing is certain, that no type of work will be satisfactory or give good results that is not in every respect first class of its kind. This is the first and essential requisite, for if the materials and workmanship are what they should be, an equally perfect dam can be built of cut stone, rubble, cyclopean rubble or concrete, nor need it be considered that this assertion at all contradicts what has been already written regarding the comparative merits of different classes of work. One class, under local circumstances, may favor good results more than another, but the result is always obtainable at the cost of more or less time, trouble and expense. The author would here reiterate what he has already said regarding quality of materials and workmanship, and the necessity that there is, that both should be first class. In hydraulic structures we are dealing with an inexorable natural agent that is sure to search out all defects and weak points, sooner or later, just as certainly as it seeks its own lowest level.

Generally speaking in comparatively low dams, no particular care is given to carry up the work uniformly. The pressures are relatively so insignificant that unequal loading of foundations, or settlement of work are not heeded. Indeed "leveling up" is avoided rather than otherwise, as it is considered best to keep the work somewhat ragged, so as to break beds as well as joints, as much as possible. In a very high dam however, the case is different, and it becomes advisable to carry up the work as nearly level as possible, leaving the top always rough and bristling with projecting stones. Concrete, of course, goes up level, but the different beds bond themselves so together that all division between the courses becomes obliterated.

CHAPTER VI.

ACCESSORIES OF DAMS.

THERE are two indespensable accessories to all dams, whether high or low, of
which it is now time to speak. First,
there must be an overflow, or spillway as
it is currently termed in the Croton dis-
trict, to provide for the escape of surplus
water when the dam is already full. This
must be of sufficient capacity to carry the
maximum volume of such water in times
of greatest freshets. This spillway may
be either artificial or natural. Whenever
the topography of the locality affords an
opportunity of passing the surplus water
through some natural depression situated
within the reservoir itself, this should
always be availed of. Sometimes such a
depression may be artificially created, by
lowering or possibly tunneling an interven-
ing rocky ridge, which separates the valley
forming the reservoir from some adjacent
one. It is generally worth while to be at

considerable expense to prepare such a natural overflow. Otherwise, a special constructive feature must be added to the dam, in the shape of a gap, through which the surplus water may pass, and fall over some designated portion of the face of the dam, into its own valley.

Considerable judgment is necessary in fixing the dimensions of the spillway and the proper proportion between its length and depth. Both are functions of the area of water shed lying above the dam. Knowing the area, A, in square miles of this water-shed, and having decided upon the probable quantity, Q, in cubic feet per second per square mile, which the water-shed may furnish in times of maximum freshets, we have for the length, L, and the depth, D, of the overflow, in feet, the two following formulæ:

$$L = 20 \sqrt{A} \dots\dots\dots\dots (16)$$

$$D = \frac{\sqrt[3]{Q^2 A}}{16} + C \dots\dots(17)$$

C being a certain additional height above the highest level of the water in the reservoir, which in a masonry dam need only be of a relatively trifling amount.

If we allow 64 cubic feet per second per square mile, which is about the smallest freshet discharge which it would be safe to count upon, (17) reduces to

$$D = \sqrt[3]{A} + C \ldots\ldots\ldots(18)$$

In the case of a masonry dam, there is less necessity for a very large spillway because even if the water should rise above its crest, and pass over its entire length, no great damage would occur, provided the foot and sides of the dam were properly protected either naturally or artificially. For an earthen dam, with masonry core, the capacity should be much more liberal, while for a dam composed wholly of earth—a kind of construction which should never be adopted—the capacity of the spillway should be still further increased, because if the water once rises over the top of such a dam, its total destruction is prompt and certain.

It may be remarked also, that the smaller the dam in relation to the water-shed, the greater necessity for a capacious spillway, because it is more frequently called into use, and must more frequently carry the full discharge of the stream. A very large reservoir formed by a high dam is much less often filled to overflowing, because it will commonly be only partly filled at the season when the freshets commence, and will therefore have a large storage capacity for the freshet flow, before beginning to run over. Indeed some large reservoirs have been built for the very purpose of protecting regions otherwise subject to inundations, by receiving and storing the water of freshets, and thus equalizing the flow of the stream on which they have been built.

The second indispensable adjunct is a waste culvert, in order to be able to draw off the water of the reservoir if so desired. This should be located at the lowest possible elevation above the bottom of the reservoir, and be large enough, if possible, to accomplish its purpose of emptying the reservoir, within a reasonable time. When dealing with the large volumes of stored water, and the abundant natural flow of the stream which are the accompaniments of high dams built across the valleys of powerful water courses, it will not be possible to build this culvert so large as to permit of emptying the reservoir by its aid at all times, and when it is desired to drain off the water down to the bottom, it is necessary to choose the dry season in order to accomplish it. This waste culvert may also become very useful as a means of supplementing the action of the spillway in times of extraordinary floods, when every effort is required to retard the rise of the water in the spill way.

The means of opening and closing this culvert requires careful study, for if the appliances employed refuse to work, either when open or shut, very serious consequences may ensue. The first appliance which naturally suggests itself, is a sliding sluice gate, closing or uncovering an opening made through the wall. The manufacturers of water supply fixtures, now make these gates in a very perfect manner, according to different designs, one of the best of these, being that adopted by the New Aqueduct Commission of the City of New York. The trouble with all such gates, when operated against a high head of water, is the difficulty of closing them, owing to the fact that this must be accomplished by heavy pressure upon a long and slender rod. In such cases they should be made very heavy, so that their weight may add to the pressure necessary to close them, and this adds to the effort required to raise them. With all their inherent defects, these gates constitute one of the best and simplest methods of controlling the waste culvert, and whatever combined system is adopted, they should form a part of it.

It will not be safe however, to depend upon a single piece of mechanism for this purpose. Another means of accomplishing the desired end is to build a pipe or—in the case of a large reservoir—pipes, into the body of the dam fitted with valves or stop cocks. These valves, operated from the outside of the dam are more certain and easier in their action than any sliding sluice gate can possibly be. These pipes and valves should form a part of the system, so that we find two necessary features which our system must embrace; large pipes built into the wall, with their mouths closed by sliding sluice gates, and the flow through them further controlled by valves or stop cocks. One gate and one valve are indispensible; in some cases two such valves on each pipe may be considered desirable, of which the one nearer the water is kept open and the flow regulated by the outer one. Should any accident occur to this outer one, the inner one can be closed, as well as the sluice gate, and the injured piece can be laid off until a convenient opportunity presents itself for repairs.

Besides these appliances, it is customary in a well designed piece of work to provide an entrance chamber furnished with grooves into which on an emergency "stop plank" may be placed, forming a temperary coffer dam around the mouth of the pipes, and thus affording access to them without emptying the reservoir. These are excellent expedients for dams of moderate heights, but it is evident that for great heights no system of stop planking could be made sufficiently strong and easily managed to be of use. In fact the enormous pressures under which all these appliances for draining off the water held up by very high dams must be operated, make their design a troublesome problem, taxing to the utmost the best skill of the mechanical engineer.

If the reservoir is to be used for the distribution, as well as the storage of water, it will be necessary to have additional appliances for drawing off the daily supply. These will be similar in kind to, but generally on a smaller scale than those already mentioned in connection with the discharge culvert. Frequently, also, sets of screens will be required, to prevent the entrance of floating objects into the distribution pipes. It will generally be desired also to have the system so arranged as to be able to draw the water from different levels, according to its conditions regarding sediment, surface impurities, etc.

Probably the best way to accomplish all the desired requirements is to place a rectangular chamber, running the whole height of the dam, directly in front of the mouth of the distributing pipe or pipes, one side of the chamber being formed by the back of the dam. The opposite side to this one, will have openings at different heights, closed by sliding sluice gates operated by capstans at the top. Inside of this chamber will be the grooves destined to receive the screens and stop plank if these are used. The mouth of the pipes will also be closed by sluice gates, and the flow of water further controlled by means of valves placed on the pipes in a gate house situated outside of the dam.

The designing of such a system when needed for a dam of only moderate height presents no particular difficulty. When however it is a question of adapting it to a very high dam, great care is necessary to ensure a system which will work satisfactorily under the greatly increased pressures. Naturally, the designing engineer will in such cases seek for precedents in existing works, and inform himself as to how the different methods which he may study, have behaved in actual use. This is the best plan to pursue, because the question is preeminently a practical one.

One error in designing such appliances may be pointed out and warned against, and that is too great a complication in the number and character of the appliances used. While every contingency likely to occur should be forseen and provided for, it is frequently the case, particularly when great perfection is aimed at, that there is an unnecessary multiplication of parts, the result of which is, that while remote contingencies are provided for, it is at the expense of the ease and certainty with which the daily routine of the distribution is executed. It is believed that this undesirable condition of things is sometimes brought about by the successive addition, as the design is being prepared, of features not originally contemplated when the ground work of the plan was first laid down on paper. Generally if, when the design is made, a careful study of it reveals the fact that certain probable contingencies, either of breakage of appliances or desired control of the flow of water have been overlooked, it will be better to reconsider the whole design, and see if, instead of adding some new feature or part, those already contemplated cannot be so modified as to embrace the additional anticipated duty or requirement. It is very rare that an afterthought can be made to harmonize with an original conception, and it is better to tear up a dozen imperfect paper designs, than to perpetuate an awkward contrivance in stone and mortar. Every additional part beyond what is strictly necessary, involves an additional opportunity for something to break or get out of order, and sometimes the safeguard which was intended to avert an interruption in the working of the system is itself the very cause of the interruption taking place.

It is in these features that the practical skill of the designer is shown in bold relief. A simple design, accomplishing the desired result with a few strong and

easily handled pieces, proclaims the skill-
ful engineer much more than an intricate
system of appliances, no matter how in-
genious it may be.

THE END.

www.ingramcontent.com/pod-product-compliance
Lightning Source LLC
Chambersburg PA
CBHW021420090426
42742CB00009B/1194